NO AR

T5-AQQ-438

Jobs in SUSTAINABLE AGRICULTURE

ROSEN
PUBLISHING®
New York

PAULA JOHANSON

*For Frank, who told me to write for children because
I told them stories in the field as we worked together*

Published in 2010 by The Rosen Publishing Group, Inc.
29 East 21st Street, New York, NY 10010

First Edition

Library of Congress Cataloging-in-Publication Data

Johanson, Paula.
Jobs in sustainable agriculture / Paula Johanson.—1st ed.
    p. cm.—(Green careers)
Includes bibliographical references and index.
ISBN 978-1-4358-3568-9 (library binding)
1. Sustainable agriculture—Vocational guidance. 2. Agriculture—
Vocational guidance. I. Title. II. Series: Green careers.
S494.5.S86J64 2010
630.23—dc22

                                                        2009014426

On the cover: A student at the Dickinson College Farm in Boiling Springs, Pennsylvania, feeds chickens.

On the title page: Left: A technician *(left)* and a farm manager record measurements so that they can estimate forage yields in a pasture at the Virginia Tech Shenandoah Agricultural Research and Extension Center. Right: An immunologist *(foreground)* and a veterinarian collect and label eggs that will be tested for the *Salmonella enteritidis* bacterium.

CONTENTS

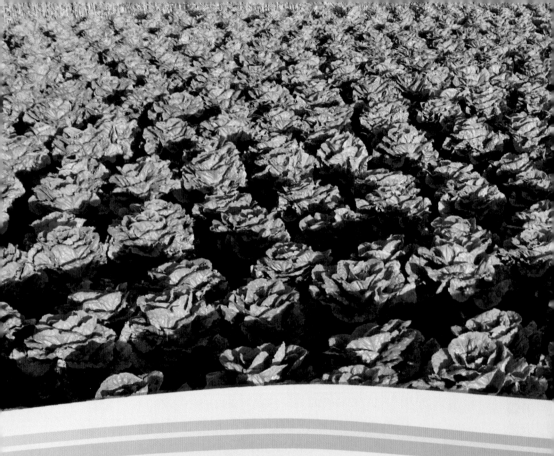

Introduction

Farmers are not the only people involved in agriculture. Anyone who wears clothes, uses wood or leather, or simply eats food is part of agriculture. "Eating is an agricultural act," says poet Wendell Berry, who is also a Kentucky farmer. "Eating ends the annual drama of the food economy that begins with planting and birth."

For thousands of years, sustainable agriculture was an idea that didn't need a special name. People raised plants and animals using traditional methods, which either sustained the natural resources of a place or encouraged people to move their work from one place to another over a wide area so that the land could recover and be used again.

4

But during the twentieth century, the farming methods in many countries changed with the use of petroleum-based fuels and fertilizers. "Agriculture has become agribusiness, and care of the land often comes a poor second to maximizing output," Janice Rudlin writes in her book *Land Abuse and Soil Erosion*. Traditional farming methods were lost or suppressed in the drive to be modern and efficient at producing a product for sale. Communities suffered when profits and benefits left the area where the agricultural products were grown.

People are always looking for ways to do their work effectively, and they've learned that the newer agricultural methods are not sustainable. The soil loses its fertility, the plants and animals lose genetic diversity, and the environment is less able to sustain people doing a variety of work. Even the agricultural products for sale are not sustainable, either. Huge fields of a single crop of plants all genetically alike are vulnerable to disease and weather changes. Clear-cut forests lose topsoil and undergrowth, so the new growth of trees never becomes as useful and healthy. Thousands of animals bred from a single father suffer from being crowded into feedlots or stacked in cages in a barn, and their dung taints the groundwater for miles around. "An interest in agriculture, its evolution, and its changes should be a priority for everyone who eats," writes Carlo Petrini in his book *Slow Food*. "But for many people that is not the case, and as well as harming ourselves and paying for it in loss of flavor and poverty of diet, we automatically become accomplices of the devastation that is wrought upon the earth by the spread of unsustainable agricultural methods."

Instead of these bad choices, many people prefer to work in sustainable agriculture. Many career choices are available, including farm manager, greenhouse worker, supermarket produce manager, research scientist, beekeeper, environmental engineer, allotment garden coordinator, and forester. For every person who raises plants and animals, there are several who work in closely related services essential for agriculture, supporting farmers in turn. These interdependent careers sustain not only production but also a web of interconnected services. Petrini considers agriculture and ecology to be one discipline, saying, "Anyone who tills the land and

6

raises livestock works with nature and must not exploit and kill it."

This sense of working with nature, not battling it, gives dignity to careers in sustainable agriculture—a dignity that rewards workers and business owners who value the environment and community as much as the dollars earned selling a product. "Farming provides a dignified way for people to access food," says Mary Holmes, of the Urban Aboriginal Community Kitchen Garden Project, a Vancouver, Canada, gardening project that brings together people tending community vegetables and fruit. A common quality to most careers in sustainable agriculture is the sense of working together as part of a grand, integrated network, where humans have a role in the food web of the natural world.

Chapter One

Farm Labor in Sustainable Agriculture

Farm workers do the hands-on work of raising plants and animals. There are many options for farming careers. Although the most well-known career in sustainable agriculture is that of organic farmer, there is a great deal of different daily experience between harvesting 640 acres (259 hectares) of field peas and operating tree farms.

There are other plant-based careers, such as tree surgeon, organic horticulturalist, and landscaper. "Horticulture . . . covers all commercial and scientific enterprises involving

Like many organic farmers surveyed across the United States, these farmers find that customer demand is soaring for the organic produce they grow on their small organic farm.

8

flowers, vegetables, fruits, grass, turf management, grounds keeping, and forestry," says Blythe Carmenson in her book *Careers for Plant Lovers*. "Horticulturists work in every aspect of the plant field, from cultivation, wholesaling, and retailing to design, management, and scientific study."

Farming careers involving animals include rancher, dairy farmer, poultry farmer, and beekeeper. Each of these careers puts different demands on a person's time and skills. Small family farms may work with both plants and animals.

Common Dangers of All Farm Work

The biggest danger faced by all agricultural workers is accidental crippling injury. An injury that keeps someone from working for weeks or causes a permanent disability is something that has to be avoided each day—and sometimes every hour. These injuries occur because of heavy lifting or accidents with a tool or an animal.

A farm is a more dangerous place than an office. Farm laborers must watch out for worksite hazards such as fire, a falling stack of hay bales, and unsafe operation of machinery by themselves or others. Every year in the United States and Canada, there are injuries and deaths among agricultural workers and their children at their worksites.

Career Options as an Organic Farmer

Organic farming is a good career for someone who enjoys plants and working outdoors, either alone or with a few

9

people. An organic farm may be a large operation or a very small garden. Three out of four organic farmers are self-employed, market gardeners who work on 1 acre (0.4 ha) to 160 acres (64.75 ha) or more. Others are farm managers for absentee owners or older relatives, or they are employed as farm managers or supervisors of corporations maintaining several large farming operations.

Starting a career as a self-employed organic farmer involves a large investment in land, but often a farmer will be able to live and raise a family there. It can take twenty years to pay off a mortgage, especially when the debt load increases when buying necessary equipment such as a tractor and truck. Some people begin by using a few acres of fallow land on a relative's acreage. It's not essential to own land, but long-term rental agreements are better than month-by-month arrangements.

Most organic farmers learned to do the work while growing up on a family farm. There are also hands-on learning opportunities where young people can volunteer on farms for credit in high school or community college courses in biology, ecology, environmental studies, and more. Across the country, programs get young people working on farms, including the programs at Green Mountain College in Poultney, Vermont, and Santa Rosa Junior College in Santa Rosa, California. There are programs in organic agriculture at colleges that take a year or eighteen months to complete. It's possible to get a bachelor's or master's degree in sustainable agriculture, but it's not necessary. Ph.D. programs are available for those who are interested in teaching agricultural studies. Farmers who want to have their businesses certified organic may take a few one-day seminars or weekend courses in fruit

At the Dickinson College farm in Pennsylvania, student interns learn composting and other farm skills while growing produce for the college dining hall and the local farmers' market.

tree pruning, intensive gardening, or tomato production. Other courses are very useful, training farmers in computer record keeping and Internet research, business planning, payroll accounting, construction, small engine repair, animal care, food safety, and marketing. A farmer needs these business management skills as well as technological training. He or she also needs a healthy, strong body and an alert mind.

Income for a self-employed farmer arrives only when the produce is sold, usually during the summer and early fall months. Produce raised using organic methods sells for up to twice the price of ordinary produce. Certified organic

Hard Work, Long Hours

Agricultural work is hard work, driven by weather and the seasons. Some workdays start before dawn and end long after dark, often in uncomfortably cold or hot weather. Heavy lifting and repetitive tasks demand both strength and endurance. More delicate tasks require quick and strong hands. Operating machinery can be tedious. A worker must be able to finish a task and then turn to another necessary task. "Put another season's promise in the ground," sings songwriter Stan Rogers. Agricultural workers get tired and dirty. It's very satisfying, though, when all that hard work pays off in acres of healthy plants or flocks and herds of robust animals.

produce sells for up to three times the price of most produce. Gross and net profits vary widely.

Produce can be sold from a booth at the farm gate, at a good location on a major road, or near a city. Produce can also be sold at farmers' markets (or greenmarkets) from tables or the back of a truck. A few farmers do both farm gate sales and markets. While one person might be satisfied earning $1,000 from a large backyard garden, most farmers are looking to earn a living. Organic farmers can earn from $10,000 to $20,000 per year working alone, or $20,000 to $60,000 per year with two assistants. The U.S. Bureau of Labor Statistics offers current salary information for agricultural jobs (see the For More Information section at the back of this book to find out how to access this information). Large farm operations require more workers and more machinery; net profits can be much less than a third of gross profit from sales.

12

Farmers can also sell their entire crop for wholesale organic prices to a distributor. This is very convenient for farmers who don't want to sell retail or don't enjoy contact with a series of customers. Organic farmers must cheerfully answer questions repeatedly about their organic certification and organic methods. In addition, customers usually ask questions that are along the lines of "Why is this bean purple?" or "How can I cook with oregano?"

Doing a variety of tasks with heavy objects, machinery, and tools means that farmers must be very careful to avoid injury. After two or three years, a farmer may have a repetitive strain injury (RSI), such as back pain, carpal

Farmers plan ahead so that when a crop ripens, the workers and equipment are ready. It's a lot of work to harvest the produce, but it is also very satisfying.

13

tunnel syndrome, or a bad knee. After ten years, it is rare for farmers not to have RSIs. Everyday aches from working hard build strength and endurance and promote general good health.

Agricultural Laborers

There are many laborers in sustainable agriculture: tree planters, farm laborers, and migrant workers. Being an agricultural laborer is not a lifelong career goal for most young people. But many of these workers are young people, working on farms with their families or planting trees in the forests. In rural areas, seasonal work may be the only nearby employment opportunity.

Many farm workers return to school in the fall and winter. A summer job as an agricultural worker can allow someone to earn money to pay tuition at a community college. Students who grew up in cities sometimes enjoy working vacations that are very different from city life. Working for a summer or a year is a very good way to know if one wants to have a career doing agricultural work.

There are jobs on farms almost everywhere in North America. Farmers have a great deal of trouble finding enough local people who are willing to do farm work. That is why crews of migrant workers travel hundreds of miles to harvest one ripe crop and then another.

There are also many opportunities for other agricultural work. Ranches and large animal farms need workers temporarily when selling animals. Reforestation companies have noticed that tree planters have a high turnover rate— not everyone enjoys camping and planting trees all

The Hoopa Valley Indian Reservation in California owns and operates its own greenhouse business. Members of the tribe care for young trees of many kinds that will be planted on tree farms.

summer. Every summer, two out of three tree planters are planting for the first time. No experience or education is necessary for this kind of work.

The bad news is, the pay is low. Near a small town like Logan, North Dakota, a farm laborer may be paid from $6 to $11 per hour, with migrant workers being at the bottom of the scale. Farm laborers are paid every week or every two weeks during the summer. Migrant workers are paid by the volume of crops that they pick. Tree planters are paid by the number of bags of seedlings that they plant instead of the number of hours worked. A strong and

15

quick worker may be very successful. Agricultural workers must be just as careful about injury as a farmer, but they are a little less likely to have an RSI.

Jobs on farms can be found at the local employment office, by asking the district extension service agent (also known as the district home economist), or by applying at a farm. There is volunteer work with programs such as World Wide Opportunities on Organic Farms (WWOOF), the New England Small Farm Institute (NESFI), and North East Workers on Organic Farms (NEWOOF), which match volunteers with host families. There, organic farmers give a volunteer food and accommodation in exchange for a prearranged number of work hours. This is a good way to have a working vacation that lasts for months.

Living in Glass Houses

Greenhouse growers are farmers who grow plants in greenhouses instead of in open fields or who start trays of seedlings in greenhouses for transplanting into fields. The plants can be grown in topsoil, a soil-less mix including peat, or rarely, hydroponically (using mineral nutrient solutions without soil). A greenhouse extends the growing season in a cold climate. Gardening indoors is a good way to increase the amount of produce that is grown on a small property.

Most greenhouse growers are far more likely than other farmers to take at least some postsecondary courses. There are short courses in greenhouse gardening at community colleges. And one can earn one-year and two-year certificates or degrees in agricultural studies at universities. The

This greenhouse grower is part of Vermont's Agency of Agriculture program that introduces farmers and local distributors, encouraging effective local marketing of produce.

learning goes on for years, as greenhouse growers continue to become educated about horticultural methods, use of pesticides, pruning, propagation, and composting.

Maintaining a greenhouse is more expensive than growing field crops, especially when the building is serviced with electricity, water, a telephone, and sometimes even heat. These services mean that greenhouse gardens are at risk for fire damage and temperature extremes. A greenhouse grower has higher expenses than a farmer using fields the same size and a much higher gross income. Usually, the net profit will be higher as well. Organic growers can earn a net profit of $10,000 to $20,000 per year working alone, or $20,000 to $60,000 per year with two assistants. A greenhouse worker is likely to be

Which Came First—the Chicken or the Egg?

Organic poultry farmers know the answer to the old joke. For farmers who sell eggs, the birds come first. For farmers who hatch eggs and sell birds, the eggs come first. Organic poultry farmers keep busy hauling straw bales and sacks of grain, driving small tractors, and moving birds within a barn or outdoors for grazing. Handling dozens or hundreds of small, affordable animals is easier in many ways than handling large animals. There's a reason that meat and eggs taste better from poultry allowed free range in a farmyard, and it's more than adding crushed oyster shells to their feed to make their bones and eggshells strong. Sustainable farms meet the needs of the poultry, including room for exercise and access to green plants, sunshine, and dirt.

paid a dollar or two more an hour than field workers on nearby farms.

Careers Raising Animals

A career raising animals is a good choice for someone who likes animals and can plan yearly budgets. The farmer earns money only when customers buy the animals or animal products, such as eggs, milk, and honey. Sales may happen once each year, particularly for large animals, or every couple of months for poultry and other small animals.

An owner may hire a farm manager, as well as workers. Three out of four farmers are self-employed and might need to hire only one or two workers.

Most farmers who raise animals learn from growing up on their family farms. At agricultural colleges, there are programs in animal husbandry, which take a year or eighteen months to complete. Farmers who want to live a little "greener" than the usual North American farmer, or raise certified organic animals, may take a few seminars or short courses through colleges and the district extension service.

Self-employed animal farmers make a large investment in land, livestock, and equipment. In some areas, dairy and egg farmers need to buy a "quota," a onetime payment of thousands of dollars to the local marketing board to be able to sell a certain amount of milk or eggs each day to the board. This quota is a business asset that may be sold later. It can take twenty years or more to pay off a mortgage. There are collective farms, with five to fifty adults, that have paid off mortgages in five years.

19

Some ranchers begin by using a few acres of fallow land on a relative's acreage. Beekeepers put hives on other people's farms and give a share of the honey in trade. Owning the land is not required, but long-term rental agreements for rangeland or raising crops for feed are better than month-by-month arrangements so that organic methods can be used and appropriate shelter or fencing can be built.

There is work to do every day of the year. It's hard to leave the farm for a vacation, or a weekend, or for more than four or five hours. People working with animals face

Bees are essential for pollinating almost all plant food crops. Commercial beekeepers can manage hundreds of beehives or bee colonies daily, often checking bees for parasites, viruses, and other intruders.

the usual farm dangers of RSIs and other injuries, and the chance of catching a disease from the animals.

There are numerous specialized tasks. Poultry farmers, for example, have a choice of several products to sell. The smallest farms may have a variety of birds and sell both eggs and packaged meat, but larger operations usually focus on one type of bird and one product.

A farmer raising cattle or sheep gets little sleep when the young are born in the spring. The new mother animals and young may need assistance or a veterinarian. Some young animals are handfed. All the animals will be

This organic hog farm in Iowa allows the pigs more room to walk around and interact than in a crowded barn. The animals don't need antibiotics in their food of mixed grains.

handled a couple of times each year, to trim their hooves and treat small injuries.

Raising cruelty-free or certified organic animals doesn't have many extra monetary costs compared with unsustainable methods, except for the examination fees to be certified organic. Animal shelters are less crowded. Organic feed and pasture take work but usually don't cost more. Farmers spend more hands-on time with the animals instead of just putting antibiotics and hormones in their feed. Organic animal products sell for two or three times the price of products that are not certified organic. The Bureau of Labor Statistics and other references suggest that organic animal farmers and beekeepers can earn $10,000 to $20,000 per year working alone, or $20,000 to $60,000 per year with two assistants. Large farms require more workers and more machinery, which means that the net profit after expenses can be much less than a third of the gross profit.

Chapter Two

Skilled Labor Supporting Sustainable Agriculture

Professor Monkombu Sambasivan Swaminathan, winner of the 1987 World Food Prize for leading the introduction of high-yielding wheat and rice varieties to India's farmers, says, "Agriculture must help produce not only more food, but also more income and livelihood opportunities." Many livelihood opportunities are in careers in skilled labor, which are necessary to support agriculture.

Skilled laborers do hands-on work, processing plants and animals into products and marketing these products.

At an Ohio research site, an agricultural engineer, a plant pathologist, a university extension horticulturist, and a technician are working together, collecting water samples to analyze the nutrients and pesticide leaching.

23

These careers in skilled labor include distributors, slaughterhouse workers, truckers, mechanics, agricultural technicians, managers, and grocery clerks in organic food stores. Although still reliant on production from farms, these careers involve daily work that is less vulnerable to having a year's income ruined by a spell of bad weather.

Watching the Clock

The work of skilled laborers is affected by the seasons and the weather, but it's driven by the clock as well. Skilled laborers must show up for work on time and work efficiently because any delays in their work affect more than just a few people. Skilled laborers make deliveries on time, make effective use of tools and machinery, and make special efforts to have products available or extra work done when seasonal factors and weather affect local working conditions.

Agricultural Technician

When research scientists announced discoveries about the dangers of genetically modified organisms (GMOs), they didn't lift every sack of canola grain studied or peer at every sample of tobacco under a microscope. The day-to-day work of agricultural research and record keeping is done by agricultural technicians.

Most agricultural technicians are hired to work for government agencies, such as the U.S. Department of Agriculture (USDA), or a university or college research program. Often, private companies rely on agricultural technicians to ensure and maintain the clean operation of

Getting Technical About Food

After eighteen months of training at a technical college, Eric was a certified agricultural technician. His starting wage at a research center in a nearby town was twice what he used to earn working in his parents' fields picking beets and potatoes. Eric still handled vegetables, taking samples and helping his supervisor gather information on methods to store each kind of produce for the center's scientists. He preferred the time that he spent analyzing the samples, maintaining the storage facilities, and keeping the computer records. Precise temperatures, measured humidity, and modern lab equipment—it seemed that the center had everything necessary to measure and understand anything about storing produce. But when new equipment was needed to store sugar beets, Eric helped his supervisor design and weld the hopper and bins. He began to realize the responsibility of learning how a harvest could be kept from spoiling before being processed.

a warehouse, factory, or packaging plant. Like most farm workers, agricultural technicians need to be concerned about avoiding physical injuries, as well as being sure to operate machinery safely and keep accurate records.

It is possible to be hired with only a high school diploma and a year of experience in agriculture or animal science, but most employers insist on formal education in the agricultural sciences. By earning an agricultural technician or technologist certificate from a community college or technical institute, the worker will be trained in horticultural management practices, computer record keeping, fieldwork, laboratory analysis, writing reports, operating equipment and motor vehicles, and other appropriate skills. It helps to be able-bodied, as agricultural technicians must be careful to avoid injuries from heavy lifting or repeated motions. Life experience growing up on a farm or working in grocery stores is an asset.

The starting wage is $18,000 to $22,000 per year. With years of experience and further training, a technician may be promoted to a supervisory role and challenging projects that pay $36,000 or more per year.

Slaughterhouse Worker

Few people would ever choose to be a slaughterhouse worker. Those who do this kind of work usually have no skills and no other opportunity for work where they live. The pay is low, and in many cases, it is below subsistence level or minimum wage. Few slaughterhouse workers belong to unions or receive benefits.

Unlike most other agricultural careers, the dangers of working in slaughterhouses have increased with modern technology. During the past decade, conditions for animals at slaughterhouses have improved in some ways, where animal cruelty laws are enforced. "Conditions for slaughterhouse workers, however, have deteriorated," writes Erik Marcus in his book *Vegan*. "Over 70 percent of cattle raised in the United States end up in slaughterhouses owned by ConAgra, Excel, or IBP [Iowa Beef Packers]." As the "Big Three" slaughter companies buy smaller meatpackers and drive others out of business, Marcus noted that working conditions in slaughterhouses become more hazardous each year.

A local slaughterhouse may specialize in handling beef, or it may process meat from many kinds of animals, including sheep, elk, and buffalo, as well as poultry from ducks to ostriches.

There are some small and medium-sized slaughterhouses, privately owned, which are much safer and cleaner workplaces. Each of these slaughterhouses is usually one of a handful of employment opportunities within 100 miles (160 kilometers), providing local meatpacking and inspection services to organic farmers who would otherwise ship their animals more than a hundred miles to market.

At smaller slaughterhouses, workers take the time needed to avoid injury. Some small facilities rotate duties so that instead of just killing and cleaning, workers can learn the skills of meat cutting. Animals are treated with more care and consideration, and they suffer less from heat or cold, thirst, crowding, and injury. It's easier to keep a smaller slaughterhouse clean and keep pollution out of local groundwater. The meat products are inspected more thoroughly than in a large meatpacking plant, and so the products are cleaner and safer.

Kosher slaughterhouses meet Jewish cultural standards for cleanliness, and halal businesses meet Muslim cultural standards for cleanliness. These businesses usually provide better working conditions than most large slaughterhouses.

Mechanic

The work of a good mechanic is essential to sustainable agriculture. Without access to mechanics, farms can use no tools that are more complex than a hoe or a single plow pulled by a horse or ox. The tilling tools, multiple plows, and harvesting equipment pulled by teams of horses or by tractors all need to be maintained and repaired, as do tractors and trucks. Mechanics must be careful to avoid injury from the heavy machinery and electrical tools that they use every day.

Many farmers learn to do some of their own repairs. In an extended family business, one member will sometimes become a mechanic because working with machinery interests him or her more than working with plants or animals. More typically, a mechanic will start a small business or be hired by a repair shop run by a car dealership or farm equipment retailer. Mechanics can find enough work to keep them as busy as they choose to be, anywhere in the world. The average yearly income for an auto mechanic is $30,000. And an experienced mechanic of agricultural equipment may earn $40,000 per year.

A mechanic can be trained and certified in a year with an apprenticeship or two years with a journeyman's certificate through a trades program at a community college. Some mechanics specialize in one type of machinery, while others get a variety of training taking extra courses in heavy equipment, welding, and hydraulics. As new technology is invented, taking upgrade courses every few years is a good idea.

Adding Value to Products

There's a lot of appeal in taking basic agricultural products and turning them into processed goods that can be sold for a much higher price. Instead of selling tomatoes at the same price as other growers, a person might sell jars of salsa. The product has had value added. The products might be sold to a wholesaler or directly to customers at the farm gate or markets.

Community colleges and agricultural colleges provide many opportunities to learn business administration skills. These courses help people improve their small business planning and operation.

This work is good for someone who likes being self-employed and likes taking chances in developing new products. For example, a farmer with more yellow peas than he could sell tried to invent a recipe for hummus. Instead, he ended up creating NoNuts Golden Peabutter, marketed as an alternative to peanut butter. Value-added products are an opportunity to develop local businesses that support sustainable agriculture.

When starting a small business, take business education courses about computers, budget planning, and record keeping. Community centers offer non-credit "value added" courses. There are value-added initiatives and assistance programs offering short courses and one-day seminars at many state university extension or other agriculture programs, USDA offices, and rural development centers. There is no certificate available for value-added product programs at colleges, but there are degrees in business administration at universities. A master's degree in business would be useful when a business has expanded to have several employees and a gross income of hundreds of thousands of dollars.

Employees may be hired to work in sales, or shipping and receiving, or in the office. An employee who wants to do more than moving heavy items around a warehouse needs to finish high school and take at least a course or two in business education.

There are many places to find information. District extension agents know about local courses. Brian Nummer, of the University of Georgia's National Center for Home Food Preservation, suggests, "An Internet search using terms such as 'valued added agriculture' generates a list of Web sites." Also check out the Cooperative State

Research, Education, and Extension Service (an agency within the USDA) and USDA service centers.

The biggest dangers of value-added products are the breaking of regulations and the harming of customers with food that has not been prepared, preserved, or labeled properly. Local governments may insist that food for sale be prepared in an inspected kitchen, whether it's private or is in a restaurant or school. State or local governments may have an incubator kitchen program, which offers shared workspace, cooking equipment, and business advice for small entrepreneurs. Or there is a test

When a home kitchen is too small and a factory is too big, business development facilities make commercial kitchens, also known as kitchen incubators, available. There, small- and medium-sized businesses produce and package locally made food products.

kitchen at a state university's food science department. These kitchens can be rented by a producer to prepare, for example, hundreds of pies and freeze them for sale to a wholesaler or at a farmers' market.

Part-time summer work to earn pocket money could result in as little as $500. But working full-time selling products year-round to a wholesaler, a person can earn $50,000 per year. It's not unusual for a person with a simple product to earn $6,000 to $12,000 per summer, selling at a weekly farmers' market in a large city.

Transportation

Truckers are essential to sustainable agriculture. If an agricultural product is not used where it was raised, it must be transported to other people. The majority of agricultural transportation and organic foods distribution in the United States and Canada is done with motor vehicles.

Truck engines can be powered by biodiesel instead of gasoline. A trucking company can operate vehicles of various sizes, including bicycles with cargo trailers for city deliveries, using the appropriate size of container for each delivery. Trips can be scheduled for maximum efficiency. Some trucking companies use large packing crates that are transferred to freight trains or container ships, which are more fuel-efficient than trucks for moving produce thousands of miles. Other companies offset pollution from their vehicles by sponsoring tree planting and other sustainable environmental programs.

Many workers drive small, medium, or heavy trucks for transport companies or as owner-operators. Some companies offer benefits, commission, profit sharing, or bonuses.

Some truck drivers operate a family business making local deliveries of their own farm products. Others work for local cooperatives, and many work for regional branches of larger transportation companies.

The starting wage may be as low as $8 per hour, but the highest-paid truckers can earn more than $50 per hour. The average pay for grocery truck drivers is between $12 and $20 per hour. After taking driver training courses and examinations in the North American Industry Classification System (NAICS), Hazardous Waste Operations and Emergency Response (HAZWOPER), Workplace Hazardous Materials Information System (WHMIS), and Material Safety Data Sheets (MSDS), a truck driver may work for twenty or thirty years or more.

Some drivers develop back problems from sitting for such long hours and loading or unloading deliveries. The

Profiting from Sheep's Wool

Most North American sheep farms earn income from selling lambs for meat. The low price for wool isn't worth the wages for someone to stuff shorn fleece into a sack to be sold to a large wool mill. But Marie was a knitter and craftsperson, as well as a sheep farmer. She found a small wool mill in a nearby town that washed fleece using biodegradable soap and carded wool into quilt batting and yarn. After paying to have some fleece processed, Marie was able to sell quilt batting, yarn, and knitted toys at farmers' markets for a profit.

biggest danger is vehicle collisions, which can be fatal or disabling. Even minor collisions can result in dangerous spills of cargo—vinegar is an organic weed killer, but spills must be handled safely! Many truck drivers are union members of the Teamsters (formerly known as the International Brotherhood of Teamsters, a labor union in the United States and Canada that represents people in all aspects of transportation) or other unions.

An afternoon seminar taught by a sustainability coordinator can give a group of truckers the information they need to start making good choices in supporting sustainable agriculture. Meetings that take place just a few times each year can provide truckers and producers with opportunities to discuss their transportation choices.

Chapter Three

Professional Work Supporting Sustainable Agriculture

There are many careers in sustainable agriculture for professional-level workers. These careers include environmental lawyers, architectural designers, water and soil quality analysts, farm injury physiotherapists, and instructors, technicians, and technologists of many kinds.

"People who protect the natural world from misuse and who oversee the wise management of natural resources"

These high school students are learning from their agricultural education instructor about measuring groundwater depth, to observe the influence of irrigation on how much water is available in the soil.

36

are what Malcolm L. Hunter Jr., professor of wildlife ecology at the University of Maine, defines as conservation professionals in his book *Saving the Earth as a Career: Advice on Becoming a Conservation Professional.* "The umbrella stretches from those who actively support careful use of natural resources (for example, most foresters, game managers, fisheries managers, and range managers) to preservationists who strive to protect nature from human intrusions." Many of these professions support sustainable agriculture as well as the environment at large. A lawyer specializing in environmental law can work with a community of farmers to represent the environmental interests of an area in court cases. Even an insurance adjuster in an agricultural community can support sustainable agriculture.

"We are all learning this new lifestyle together," says municipal committee worker Barbara Kessler, a journalist and editor of the Web site GreenRightNow.com. "We can all be green and start to make a difference in the world—not at some future point when it's convenient—but right now."

Environmental Sampling Technician

It is essential for organic farmers to know when a nearby oil well is contaminating the soil or if gasoline leaking from an old gas station is in the groundwater. Environmental sampling technicians gather water and soil samples to be tested in a laboratory. These technicians usually work for an environmental agency under the direction of an environmental engineer, who plans the projects and assigns technicians to much of the work on the ground. This is a good career for men and women who enjoy working outdoors and want to work in the sciences.

37

Many technicians belong to professional associations or unions representing employees of the company where they work. After a few short courses in NAICS, HAZWOPER, WHMIS, and MSDS, an environmental sampling technician just starting out might earn $20,000 to $24,000 per year. An experienced technician with a one-year or two-year certificate from a community college can earn from $30,000 to $44,000 per year. This career has many opportunities for promotions and upgrading skills.

On the Spot

Pollution is cleaned up by people doing practical, hands-on work. Environmental sampling technicians are on the frontline in defending sustainable agriculture from pollution, identifying tainted sites, and taking water and soil samples to prove the type and location of contaminants. On short winter days in Alaska, techs are in the field even when it's -22 degrees Fahrenheit (-30 degrees Celsius). Scorching summer days are hard, too, and sometimes sunscreen and bottles of water aren't enough to make a long day outdoors comfortable. Then when the technicians are back in the environmental company's office, there are reports to write and records to file. This is the day-to-day work that gets toxic sites cleaned and gets the data needed to hold polluters accountable. It's very satisfying to be part of the pollution solution. Some techs find that if it's hard to be promoted within one company, they are "headhunted" by another company. During ten or twenty years of environmental sampling work, a technician might work for three or more companies, or stay within one department of a government agency.

The hazards of the job include exposure to petrochemicals derived from fossil fuels. A technician will routinely be splashed when taking samples of groundwater contaminated with fossil fuel products. There are volatile petrochemicals in the air at some work sites. Filtered breathing masks do not always offer enough protection against the more dangerous chemicals. Environmental sampling technicians have to be very careful to minimize their exposure to fossil fuel products in the field, or their health may be affected. Working outdoors much of the time promotes physical fitness and general good health.

Equipment Design Technologist

Someone invented a cattle squeeze (a chute for restraining cattle) and every other piece of equipment that is used by farmers. Agricultural equipment design technologists are behind the most recent innovations. These technologists design the right tools for the right person to do the right task, and often that person has a physical disability. Since 1999, the fastest-expanding field of farm equipment and machinery attachments is designed for farmers who are paraplegic.

The technologists who design agricultural equipment think about the users, not just efficient machinery. When comfortable seats were installed on tractors and combine harvesters, the number of farmers with chronic back pain did not decrease. As always, farmers worked until they hurt too much to do any more. But with comfortable seats, they could work many more hours in a day.

After buying land, the biggest expense in farming is agricultural equipment. A tractor isn't fast or fashionable like a Ferrari sports car, but a tractor can cost twice as much or more.

The certificate to become a design technologist takes two years to earn. The big danger in this job is letting one's skills become obsolete. Upgrade computer systems regularly, and take new courses every couple of years when new models of equipment are released on the market.

Many design technologists work for agricultural equipment companies, but others are independent entrepreneurs who make prototypes to sell to the highest bidder.

The income for these technologists varies, from $30,000 per year when starting out to $125,000 per year for the most successful inventors who register patented designs.

Pulling Together All the Plans

Sustainability coordinators work on the campus of a university or college, or they work for a corporation. Their duties change with new seasons and new projects in which they integrate sustainability into many activities of the campus or corporation. Resource security and local production of food may be important for one coordinator, while another spends weeks focused on food-related teaching and research projects.

Research assistants from the University of California Cooperative Extension are working with farmers to help them develop sustainable farming methods to reduce the use of pesticides in their fields.

"There's a lot happening in sustainability, and there are a lot of people working on it, but most of them also have full-time jobs," says Nurit Katz, the first sustainability coordinator for the University of California, Los Angeles (UCLA). After getting an M.B.A. and a master of public policy degree, Katz developed and completed a four-course certificate program called Leaders in Sustainability.

While there is no professional association, some coordinators are members of unions and teaching associations. Coordinators may work full-time or part-time. The required education usually includes at least one degree in a related field, such as environmental studies or agriculture. After five years of experience, a sustainability coordinator might earn $50,000 per year. An organization may have no budget at first for a coordinator's salary. A coordinator beginning work might be told, "If you can save enough energy spending to make the equivalent of your salary, you can have a job."

Writer

While a farmer walks the walk in sustainable agriculture, writers talk the talk. Many careers in sustainable agriculture require writing skills to compose reports or apply for grants. Writers prepare reference materials, pamphlets, and books for agricultural workers. In addition, nonprofit organizations, activist groups, and the USDA all hire employees with professional writing skills.

Some journalists and columnists earn $15,000 to $40,000 per year. Other writers for government publications may earn $20,000 to $50,000 per year. Usually,

Writers support sustainable agriculture with books, articles, letters, grant applications, and more. Writing can be done almost anywhere. Pens and paper are cheap, and computers are available in public libraries and schools.

these writers have a bachelor's degree in journalism, writing, or English, or at least a two-year diploma from a community college, plus technical training in an agricultural science.

Many writers are self-employed freelancers; they are paid for each completed project. A freelance writer may have a degree, a diploma, or only a high school education and a few writing workshops under his or her belt. The major danger a writer faces is low income. Many earn as little as $5,000 per year. Some freelance writers earn $30,000 or $50,000 per year—they're usually experts doing technical writing.

Writing makes good part-time work. It can be a second career for someone who has suffered a disabling injury. Organic farmers often create recipes for customers and stories for local newspapers. They also write books on sustainable agriculture—a classic case of "write what you know."

Chapter Four

Government Work Supporting Sustainable Agriculture

Careers supporting sustainable agriculture exist in the local, state, and federal levels of government. Government employees include professionals such as public health inspectors and veterinarians. There are officials doing forestry, waterways, and farmland protection. There are also environmental engineers and cooperative

Plants grow differently in air with varying amounts of carbon dioxide. This technician is using a pressure bomb to compare the water status of plants grown in experimental tunnel greenhouses.

45

extension agents supporting farmers and skilled laborers at their work and marketing.

The work of many government employees directly supports sustainable agriculture, even if the employees' job titles do not immediately seem related. A librarian might work in the National Agricultural Library of the USDA, for instance. And the armed forces hire many environmental protection specialists. Even elected officials are part of the infrastructure that supports farming.

Research Scientist

Scientists hired by the USDA work in hundreds of laboratories across the United States. The National Research Council runs similar research centers in Canada. Some of these scientists have bachelor's degrees in science, others have master's degrees, and many have doctorates in statistics, ecology, genetics, botany, microbiology, and other areas.

Research scientists work with environmental engineers, technicians, and technologists, who gather data. Most projects involve accurate record keeping of each day's work by the scientist and several technical assistants. This is a good career for someone who loves science and thrives on being part of an organization with regulations and regular work hours.

There are many positions available globally with departments of agriculture and agricultural corporations. When starting out, a research scientist might earn only $20,000 per year as a graduate student taking a master's degree. With years of experience and a good list of publications, though, he or she can earn $50,000 to $80,000

Winning the Lottery

If looking for a career feels like entering a lottery, working at a government job feels like winning one. Reliable working conditions and standards, consistent pay scales, benefits, contracts supported by unions or professional associations—many workers prefer government jobs to similar self-employed work or private-sector jobs. Federal job benefits are listed online at http://www.usajobs.gov/ei61.asp. Some people working in the private sector perform government work on contract. Others take an official position in public service. "Starting salaries for those in private industry are generally about 20 percent higher" for professionals with degrees, says Michael Fasulo in *Careers in the Environment*. But private-sector careers are more vulnerable to business failure and economic downturns than a steady, reliable government career.

per year or more. Most research scientists belong to at least one academic and professional organization.

Who Can You Call?

Got a question about agriculture? Consult a district extension agent (also called a district home economist, agricultural agent, or county agent). This versatile government employee serves as a resource for communities and counties, helping people access information to find the answers they need. On any given workday, an agent may visit a local radio show, answer e-mail queries about food regulations, or facilitate workshops for people wanting help with product development.

Universities are more than ivory towers of knowledge. This research scientist, who is an agent of the University of Vermont Extension Service, meets with farmers to discuss various growing problems with barley, wheat, oats, and soybeans.

Education is different for each extension agent. Most have at least a bachelor's degree in agricultural studies, business management, or the sciences, complemented with experience in agriculture and small business operation. A district extension agent is a polymath, with training in multiple areas of expertise and years of experience facilitating learning for people with various needs.

The pay rate differs regionally for this position. A district extension agent working in Carson, California, might earn $65,000 to $80,000 per year, while in Butte, Montana,

A Community Resource

District extension agents are constantly busy providing information in response to mail and e-mail requests. Other requests to teach local sessions on organic farming techniques take time for preparations and community visits. Agents must be aware of local conditions. It does no good to recommend berry picking to Alaskans, for example, without cautions about bears as well as poisonous berry varieties. Linda Tannehill, district home economist for the University of Alaska, calls berry picking "a wonderful way to experience the place we live." She and her colleagues make guidebooks, identification manuals, and recipes available at extension service offices.

the average rate is around $50,000 per year. Some jurisdictions assign fewer duties and pay lower wages.

The biggest danger for these agents is when there is a health crisis in the community. If people become ill and the cause seems to be related to local agriculture, a district extension agent will make public statements in the media. The agent will certainly feel responsible for the health of everyone who follows his or her advice.

Landfill and Recycling Station Worker

When landfills and recycling stations are run with sustainable goals, the farming community benefits. A recycling station worker doesn't just handle trash. He or she also keeps track of various materials arriving at the site. Recycling goes beyond piling up bins of glass, metal, and

This USDA compost facility is located at the Beltsville Agricultural Research Center in Maryland. Composting and recycling are part of sustainable waste management for agricultural communities.

plastic. When a farmer drives in a truck filled with old appliances, wooden frames, plywood, empty oil containers, chicken feathers, and torn feedbags, the worker directs each material to its place in the waste stream. Attentive workers recycle as many materials as possible, incinerate as few as they can, and minimize the amount that is buried in landfills. Sustainable waste management supports an entire district.

Recycling station workers must be careful not to become ill from human and animal germs in the postconsumer waste. There is also a risk of being poisoned by toxic materials and petrochemicals in pesticide containers.

A city or county hires most workers at landfills and recycling stations. But some recycling stations are privately operated. Workers take courses in NAICS, HAZWOPER, WHMIS, and MSDS, and they learn to operate forklifts and tractors. Wages start at about $20,000 per year. With experience and further education in waste management, there are opportunities for promotion.

Veterinarians

Dogs and cats aren't the only animals that a veterinarian treats. Most vets work with farm animals, ensuring their health and safety. This is a particularly good career for someone who enjoys animals and has a strong interest in the sciences. The hazards of veterinary work include being injured by an animal on a farm or at a research center.

Veterinarians need a degree in veterinary medicine from a university, taking four or five years including practicum work with animals. There are other courses in alternative and complementary medicine, such as holistic medicine,

51

Veterinarians need both medical skills and the knack for handling animals confidently. This vet is taking a sample of a sheep's blood to test for genetic resistance to scrapie, a fatal disease characterized by itching, loss of muscular control, and degeneration of the nervous system.

acupuncture, chiropractic, and naturopathic medicine, that some vets believe benefit animals. Animals that need a vet's care can still be certified for organic products if the vet is careful to follow organic standards for the permitted treatment.

The Food Safety and Inspection Service division of the USDA is the largest employer of veterinarians in the United States. The Canadian Food Inspection Agency is the largest employer of vets in Canada. A vet working in a private practice might earn $34,000 to $50,000 per year, while the average for USDA vets is $72,000 per year.

Legislative Workers

Elected officials at local and federal levels create laws regulating food safety and the production and distribution of agricultural products. For every elected official, there are office staffs. Lobbyists and advisers represent the interests of private-sector industries and unions. All of these people can support sustainable agriculture if they are aware of how their actions affect sustainability.

These legislative careers are good choices for people who enjoy working in offices. The biggest danger of such work is that if it's done without understanding sustainable agriculture, the resulting decisions will be unsustainable. However, legislative workers' income is not affected by their knowledge of sustainable agriculture—if all that's being counted is dollars. For example, it's hard to count in dollars the effects of watershed protection enforcement on the health of everyone in that jurisdiction. Some legislative specialists, with the appropriate experience, can make from $90,000 to $140,000 or more working in the federal government.

"We all need to begin to pay attention to [legislation such as] the Farm Bill, working to develop farm programs that allow farmers to stay in business without falling into the trap of overproduction," said journalist Michael Pollan, author of *In Defense of Food*, when addressing the 2005 Ecological Farming Conference. "Most city people don't realize the stake they have in it . . . The Farm Bill sets out the rules of the game that everyone is playing in, whether you're an industrial or an organic farmer, whether you're eating industrially or not."

The Farm Bill, which became the Food, Conservation, and Energy Act of 2008, established new plant pest and disease management programs and provided money for farmers' markets and expanded fruit and vegetable market news reporting, among other requirements. It also increased funding for data collection on organic agriculture and to help support federal organic regulatory activities. Another bill to promote economic recovery through green jobs and infrastructure, called the Green Jobs and Infrastructure Act of 2009, is currently pending in the U.S. Congress.

Legislative professionals in sustainable agriculture, such as directors of policy, help carry out policy programs by working with business, agriculture, and government leaders to find practical ways to develop and maintain healthy ecosystems and to find environmental solutions that are economically sensible for industries as well as the public. They have experience working with regulatory agencies and understand how to build constructive and effective working relationships with scientists, government representatives, conservationists, community leaders, land owners, elected officials, and businesspeople, so that sound environmental stewardship is attained.

Chapter Five

Applying Education to Improve Careers in Sustainable Agriculture

For a lot of people, agriculture seems like simple, old-fashioned work. After all, who needs to go to college to study how to hoe beans, feed hogs, or haul logs? But the truth is, even a little education goes a long way to make sustainable agriculture effective, efficient, and profitable. New training keeps knowledge current, now that ergonomic hoes are, in fact, being designed, as

Only a short distance north of New York City is the Stone Barns Center for Food and Agriculture in Pocantico Hills, where community-based sustainable food production is demonstrated and promoted.

55

well as machinery that is safer to operate than previous equipment. And traditional seeds are now being compared to new hybrids. Education can be applied to any agricultural career.

"An education isn't how much you have committed to memory, or even how much you know," wrote philosopher Anatole France. "It's being able to differentiate between what you do know and what you don't."

Many colleges offer diplomas, certificates, and degrees in sustainable and organic agriculture. For some people, full-time study is not affordable or even locally available, but alternatives do exist. There are distance-learning courses offered by some universities and master gardening courses on weekends and evenings at community colleges.

In addition, there are free and low-cost seminars held on weekends and evenings in communities. Informal learning sessions help people learn techniques that will improve the farm work they've been doing for years. These sessions also teach city people to increase the way that their work supports sustainable agriculture. If there are none of these sessions in your community, ask for them at the community center, library, school, and the nearest USDA service center.

Who Needs Higher Education?

There are other incentives for workers to obtain post-secondary diplomas, certificates as technicians or technologists, or degrees. The certification tells potential employers that applicants are serious enough about careers in sustainable agriculture to invest time and money in appropriate training.

With the right education, it's much easier to work in fields that are challenging, interesting, and earn more money than unskilled labor. A certificate as a technologist takes less than half the time and expense to earn as a bachelor's degree, and it's appropriate for beginning work at an agricultural research station. If instead you want to work as a research scientist, you will need a master's degree or doctorate in the sciences. As Michael Fasulo points out in *Careers in the Environment*, "Obtaining a graduate degree greatly enhances job and advancement potential."

North American credentials are recognized around the world and by the Food and Agriculture Organization of the United Nations. Workers who travel abroad can find many opportunities overseas for short-term and long-term

Sharing Education

A formal education in agriculture benefits more than the student. Often in an extended family or small business, only one person will take a program at an agricultural college or become certified in a trade that is associated with agriculture. Or perhaps one worker learns greenhouse skills while another learns to maintain and repair tractors and farm machinery. The entire operation benefits. Shared knowledge spreads throughout a community. Seeing trained workers in action with unskilled workers helps young people choose a career. Moreover, unskilled workers can be inspired to get training to improve their job prospects for the future.

contracts in sustainable agriculture, working with governments and the private sector.

Organic Restaurant Manager

Restaurant managing is a risky career. In some cities, nine out of ten new restaurants close within a year. But in most communities, there are restaurants that have been in business for years.

A restaurant manager can be hired by a business owner or be a self-employed owner-operator. Some are

Shown here in her California garden, acclaimed restaurateur Alice Waters is a leading figure in the "slow food" movement. An alternative to the fast food culture, the slow food movement promotes awareness of how food is grown, prepared for meals, and enjoyed.

also the chef or chief cook. Many managers take business management and food safety courses and study cooking for a year or more so that they can be knowledgeable in their field. The latest trend that a restaurant manager needs to know about is the use of organic and locally grown foods. A restaurant manager who takes a course or two from the district extension service can learn enough about organic and local food production to increase the quality of the food served and make the business more sustainable and profitable.

Restaurants specializing in locally grown and organic foods are popular from small towns to large cities. The food tastes better than franchised fast food that is made with processed ingredients shipped for hundreds of miles. A restaurant manager familiar with the benefits of sustainable agriculture could support the "Hundred-Mile Diet" with a special three-course meal or by highlighting on the menu all the dishes made using ingredients produced within 100 miles (160 km). By purchasing from local and organic farmers, the restaurant supports local business. The restaurant also reduces its own carbon footprint when it chooses not to use ingredients shipped long distances. There are no additional hazards when operating an organic and local food restaurant.

Applied Knowledge

Even a little education has benefits. A person earning a modest living as a migrant worker has real opportunities to improve his or her skills and opportunities for promotion. A farm worker who gets certified as a pesticide applicator by taking a course from a technical college,

Applying agricultural sprays does not have to be dangerous. With a little training, workers will know how to use and apply organic sprays appropriately for their own safety and for healthy crops.

university, or government agency gains practical skills in sustainable agriculture.

Pesticide applicators are trained to make sustainable choices for controlling weeds and insects. Even certified organic farms can use certain herbicides and pesticides. Before taking an examination, the worker may take a four-day tutorial and complete home-study lessons for sixty to eighty hours. It is an effort for a migrant worker to afford such specialized training, but the modest expense will pay off for years. Workers certified as pesticide applicators are of particular interest to organic farmers who are looking to employ regular, as well as temporary, employees.

Training to apply pesticides adds no more risk to a farm worker's career. In fact, personal health and safety improves for workers who use both petrochemical and organic pesticides properly. On many farms, untrained workers with inadequate supervision routinely apply agricultural sprays. Being a certified applicator opens up opportunities for a worker to correct unsafe working conditions, moderate the appropriate use of pesticides, and protect local rivers and homes from accidental contamination.

The Meat of the Matter

Cutting and packaging meat is pretty basic work, but some meatpacking plants are more sustainable than others. This is due to the choices made by managers. Some meatpacking plant managers insist that their plants can't afford to operate as organic or cruelty-free businesses and that the effort isn't worth it. Yet others find that implementing "green" or sustainable business methods pays off in many ways. What makes the difference is the manager's knowledge of sustainable options, from biodegradable packaging materials to rotating work assignments.

Large meatpacking plants process great volumes of meat and, therefore, are at high risk for contamination with *Salmonella*, *Listeria*, and *E. coli* bacteria. A manager whose goals are sustainable knows that sustainable meatpacking plants handle a smaller volume of meat than an ordinary business and yet they earn a higher net profit. Increased cleanliness, quality inspection, and considerate animal management result in better-quality products being sold at higher prices. There are reduced absences, illnesses, and turnover among staff.

61

The manager of a meatpacking plant can earn anywhere from $30,000 to $80,000 per year, depending on training, experience, and the volume of business done by the organization. It's hard to count in dollars the value of sustainable methods like hosing out the facilities into a settling pond instead of a nearby river.

Market Management

Managing a farmers' market is a good career for someone who enjoys talking with people. A farmers' market manager is hired by the group of market vendors, by the local chamber of commerce, or by a town office. The contract does not usually offer benefits unless all employees of the town office or chamber of commerce receive benefits.

A few markets operate year-round. The work is usually seasonal and "revenue neutral"—the manager signs up enough vendors for the full season so that vendor fees add up to pay the manager's wages as well as the expense of operating the market. The pay rate can be as low as $50 per week to manage a small-town weekly summer market. Managers can earn $20,000 to $40,000 working year-round promoting a market with one hundred or more vendors, each paying $400 or more for the season.

Some managers rely on a young assistant or vendors to move tables. Vehicle traffic is a danger for all at the market. Because dog owners regularly walk their dogs through farmers' markets, managers should carry dog treats, just as animal control officers do.

Education improves the effectiveness and income of market managers. Business skills, such as record keeping

This greenmarket manager oversees the mobile farmers' market in the lower 9th Ward in New Orleans, Louisiana. In neighborhoods still recovering from the devastation of Hurricane Katrina, temporary markets give access to fresh fruit, vegetables, meat, and seafood.

and project planning, are taught in one-day seminars and through business courses held at community colleges. A smart market manager attends information sessions and seminars on organic farming, value-added products, and sustainable agriculture.

You Are What You Eat

"Eating takes place inescapably in the world; it is inescapably an agricultural act," Wendell Berry believes. "How we eat determines, to a considerable extent, how the world is used . . . To eat responsibly is to understand and enact, so far as one can, this complex relationship."

It makes sense that people interested in good food and good jobs need to have "an environmental conscience and be well informed about ecology," as Carlo Petrini writes in *Slow Food.* "At the same time, environmentalists must understand that even if you do not use chemical products, you can destroy the environment by eliminating biodiversity (such as woodland, other plants) in favor of a single variety that is produced in large quantities."

It's not enough to work hard. Agricultural workers must also work well to produce lasting benefits that sustain individuals, communities, and the environment. "Above all," Petrini adds, "those who cultivate the land must not forget taste. If a product is not good, there is no point in it being organic; if it is not good and is alien to the local culture, it may be useful in responding to an emergency, but it will not permanently solve the problem of hunger or of certain kinds of pollution."

When considering the many agricultural careers discussed here, don't lose track of farming among other work that earns a higher income. The point is to understand that farming is part of many kinds of work. These careers are not separate interests but are integrated with the single goal of sustainable agriculture.

Glossary

batting Layers or sheets of wool used for lining quilts or for stuffing or packaging.

carbon footprint The amount of carbon released into the atmosphere, usually through use of gasoline-powered vehicles, by creating and marketing a product.

cattle squeeze A stall with a set of bars that can be closed gently to hold a cow or a bull still, allowing it to be handled safely for physical care or veterinary treatment.

cruelty-free Describing traditional methods of raising animals to allow them free movement in a barn, yard, or field (free range) so that the animals don't injure each other from crowding and are not kept in wire cages stacked high in barns.

environmentally sustainable Describing methods of using natural resources that minimize pollution and maintain the ecology of an area, including recycling programs and responsible management of resources.

fallow Land that is being rested after growing crops; it may be tilled so that nothing grows, or it may be sown with cowpeas or ryegrass as a green cover to crowd out weeds before being ploughed under as green fertilizer.

greenhouse A large shed covered with sheets of glass or plastic to shelter growing plants.

HAZWOPER Hazardous Waste Operations and Emergency Response, a federal regulation.

holistic Describing a medicine or technique that tries to treat the whole mind and body.

infrastructure The system of buildings, transportation systems, water and power systems, services, and education that supports a nation.

MSDS Material Safety Data Sheets, a standard system of presenting information about materials used at work sites, usually used with WHMIS.

NAICS North American Industry Classification System, an international system for classifying industrial materials.

naturopathic Describing a treatment that avoids the use of drugs and surgery and that stresses the use of natural agents (such as air, water, and sunshine) and physical means (such as manipulation and electrical treatment).

petrochemical A chemical made from petroleum, or crude oil.

polymath A person of encyclopedic learning.

practicum A period of work for practical experience as part of an academic course; in other words, on-the-job experience in a course of study.

sustainable development Use of a resource in ways that support the local economy and maintain the local environment.

veterinarian A doctor who treats animals; some licensed vets also use alternative medicine, such as naturopathic, homeopathic, massage, or chiropractic treatments, when appropriate.

WHMIS Workplace Hazardous Materials Information System, a standard system of presenting information about hazardous materials used at work sites, usually used with MSDS.

For More Information

FarmStart
P.O. Box 1875
Guelph, ON N1H 6Z9
Canada
(519) 836-7046
Web site: http://www.farmstart.ca
FarmStart supports a new generation of farmers in
developing locally based, ecologically sound,
economically viable agricultural enterprises. It
encourages them to engage in entrepreneurial
strategies that creatively turn challenges into
opportunities.

GoodWork Canada
People and Planet
P.O. Box 21006 RPO Ottawa South
Ottawa, ON K1S 5N1
Canada
(613) 744-3392
Web site: http://www.goodworkcanada.ca
GoodWork Canada maintains listings of green jobs,
contracts, directorships, internships, and volunteer
opportunities for green businesses.

National Center for Home Food Preservation
University of Georgia
208 Hoke Smith Annex
Athens, GA 30602-4356

Web site: http://www.uga.edu/nchfp/publications/publications_usda.html

USDA publications are available online, giving step-by-step instructions for the home preparation and preservation of food. The site provides links to find state publications and local Cooperative Extension System Food Safety Offices.

National Sustainable Agriculture Information Service (ATTRA)
P.O. Box 3657
Fayetteville, AR 72702
(800) 346-9140
Web site: http://attra.ncat.org

This organization provides a directory of on-the-job learning opportunities in sustainable agriculture in the United States (and some in Canada). The organization is managed by the National Center for Appropriate Technology and offers information and technical assistance to farmers, ranchers, extension agents, educators, and others who are involved in sustainable agriculture.

North East Workers on Organic Farms (NEWOOF)
New England Small Farm Institute (NESFI)
275 Jackson Street
Belchertown, MA 01007
(413) 323-4531
Web site: http://www.smallfarm.org/newoof

The NESFI operates NEWOOF to encourage more sustainable regional agriculture. It's a nonprofit organization that provides information and training for beginning farmers and those transitioning to organic methods and sustainable agriculture.

USAJOBS
Web site: http://www.usajobs.gov
USAJOBS is the official job site of the USDA, the federal government, and the U.S. Office of Personnel Management. It's maintained as a one-stop source for federal jobs and employment information.

U.S. Bureau of Labor Statistics
2 Massachusetts Avenue NE, Room 2860
Washington, DC 20212
(202) 691-5200
Web site: http://www.bls.gov
The Bureau of Labor Statistics answers questions about employment in agricultural industries and other fields. Information and assistance is also available from regional offices and individual programs.

U.S. Department of Agriculture (USDA)
1400 Independence Avenue SW
Washington, DC 20250
Web site: http://www.usda.gov
USDA service centers are designed to be a single location where customers can access the services provided by the Farm Service Agency, the Natural Resources Conservation Service, and the Rural Development agencies.

USDA Agricultural Marketing Services (AMS)
1400 Independence Avenue SW

Washington, DC 20250

Web site: http://www.ams.usda.gov

The AMS maintains links to student career experience programs, as well as employment opportunities in dairy, fruit and vegetable, livestock and seed, and poultry programs.

USDA Economic Research Service (ERS)

1400 Independence Avenue SW

Washington, DC 20250

(202) 720-4519

Web site: http://www.ers.usda.gov

At ERS, more than three hundred economists and social scientists are partnered with coworkers from diverse professional backgrounds to research projects relevant to farm practices and agricultural economics.

U.S. Office of Personnel Management (OPM)

1900 E Street NW

Washington, DC 20415

(202) 606-1800

Web site: http://opm.gov

The OPM maintains the civilian workforce for the federal government. Its Web site has a page of links to frequently asked questions about employment, qualifications, and more.

Workaway

Web site: http://www.workaway.info

On Workaway's Web site, registered members can join international lists of volunteer workers traveling to new places and of host families and organizations. Volunteers do five hours of work per day in exchange for food and accommodation with host families.

World Wide Opportunities on Organic Farms (WWOOF)
Web site: http://www.wwoof.org
WWOOF is an international charity registered in the United Kingdom that arranges for volunteers to find placements on organic farms. Volunteers stay for a few weeks or up to several months doing work in exchange for food and shelter with a host family.

Web Sites

Due to the changing nature of Internet links, Rosen Publishing has developed an online list of Web sites related to the subject of this book. This site is updated regularly. Please use this link to access the list:

http://www.rosenlinks.com/gca/agri

For Further Reading

Camenson, Blythe. *Careers for Plant Lovers and Other Green Thumb Types*. New York, NY: McGraw-Hill, 2004.

Environmental Careers Organization. *The ECO Guide to Careers That Make a Difference: Environmental Work for a Sustainable World*. Washington, DC: Island Press, 2004.

Everett, Melissa. *Making a Living While Making a Difference: Conscious Careers in an Era of Interdependence*. Rev. ed. Gabriola Island, BC, Canada: New Society Publishers, 2007.

Fasulo, Mike, and Paul Walker. *Careers in the Environment*. New York, NY: McGraw-Hill, 2007.

Greenland, Paul R., and Anna Marie L. Sheldon. *Career Opportunities in Conservation and the Environment*. New York, NY: Ferguson/Facts On File, 2007.

Hunter, Malcolm L., Jr., David Lindenmayer, and Aram Calhoun. *Saving the Earth as a Career: Advice on Becoming a Conservation Professional*. Malden, MA: Blackwell Publishing, 2007.

McNamee, Gregory. *Careers in Renewable Energy: Get a Green Energy Job*. Masonville, CO: 2008

Miller, Louise. *Careers for Nature Lovers and Other Outdoor Types*. New York, NY: McGraw-Hill, 2008.

Shenk, Ellen. *Outdoor Careers: Exploring Occupations in Outdoor Fields*. 2nd ed. Mechanicsburg, PA: Stackpole Books, 2000.

Bibliography

Berry, Wendell. "The Pleasures of Eating." Center for Ecoliteracy. Retrieved January 14, 2009 (http://www.ecoliteracy.org/publications/rsl/wendell-berry.html).

Bowden, Rob. *Sustainable World: Food and Farming*. Farmington Hills, MI: Kidhaven Press, 2004.

Center for Food Safety and Applied Nutrition, U.S. Food and Drug Administration, and Department of Health and Human Services. "Food, Nutrition and Cosmetics Questions & Answers: Topics List, Comments and Questions." December 29, 2006. Retrieved January 14, 2009 (http://www.cfsan.fda.gov/~dms/qa-top.html).

Clemson University. "Private Pesticide Applicator Certification." Pesticide Information Program, November 16, 2005. Retrieved February 28, 2009 (http://entweb.clemson.edu/pesticid/private.htm).

Cornell University. "Pesticide Safety Education Program." Pesticide Management Education Program, 2008. Retrieved December 15, 2009 (http://pmep.cce.cornell.edu).

Cummings, Claire Hope. *Uncertain Peril: Genetic Engineering and the Future of Seeds*. Boston, MA: Beacon Press, 2008.

Eckfeldt, John. "Appropriate Technology for a Sustainable Future." Sustainable Technology Education Project. Victoria, BC, Canada: VIDEA, 2003. Retrieved

January 20, 2009 (http://www.videa.ca/index.php?pageid=2).

Forest Shop. "Woodlot Management." 2008. Retrieved January 4, 2009 (http://www.forestshop.com/woodlotman.html).

Goerz, Byron. *To Plant or Not to Plant—A Treeplanter's Guide.* Fort St. James, BC, Canada: Little Cabin Books, 1996.

Goerz, Byron. *Treeplanting: Is It For You?* Retrieved January 4, 2009 (http://www.unb.ca/bruns/9697/opinion/Issue21/treeplanting.html).

Heiman, Heide. *Working with Nature: Shifting Paradigms: The Science and Practice of Organic Horticulture.* Cowichan Station, BC, Canada: Gaia College, 2007.

Heintzman, Andrew, and Evan Solomon, eds. *Feeding the Future from Fat to Famine: How to Solve the World's Food Crises.* Toronto, ON, Canada: House of Anansi Press, 2005.

Hewitt, Allison. "New Sustainability Coordinator Looks Ahead to Year's Green Projects." UCLA Sustainability, November 21, 2008. Retrieved February 28, 2009 (http://www.sustain.ucla.edu/article.asp?parentid=1707).

Holmes, Mary E. "The Urban Aboriginal Community Kitchen Garden Project." Lemongrassmedia, September 18, 2008. Retrieved January 8, 2009 (http://ca.youtube.com/watch?v=8RSn9d9pMN8).

House of Commons Committees. "Pesticides: Making the Right Choice for the Protection of Health and the Environment." Report of the Standing Committee on Environment and Sustainable Development, May 2000. Retrieved January 7, 2009 (http://www2.parl. gc.ca/HousePublications/Publication.aspx?DocId= 1031697&Language=E&Mode=1&Parl=36&Ses=2).

International Institute for Sustainable Development. "Sustainable Agriculture and Rural Development." Retrieved January 19, 2009 (http://www.iisd.org/ ic/info/ss9507.htm).

Ivanko, John, and Lisa Kivirist. *Rural Renaissance: Renewing the Quest for the Good Life.* Gabriola Island, BC, Canada: New Society Publishers, 2004.

Johanson, Paula. "Agricultural Design Technologist." *Look Ahead, Get Ahead: Growing Career Opportunities for Technicians and Technologists.* Ottawa, ON, Canada: Canadian Technology Human Resources Board, 1999.

Kansas Department of Administration. "Personnel Services." January 13, 2009. Retrieved January 19, 2009 (http://www.da.ks.gov/ps).

Lamont, Eve. *The Fight for True Farming: Food Autonomy Versus Agribusiness.* Montreal, QC, Canada: National Film Board of Canada, 2005.

Marcus, Erik. "The Killing Business" *Vegan: The New Ethics of Eating.* Ithaca, NY: McBooks Press, 2001.

Newport News Times. "Local Logger Turned Writer Urges Middle Ground on Environment." February 27, 2004. Retrieved January 20, 2009 (http:// newportnewstimes.com/articles/2000/11/17/ community/community-07.txt).

Nummer, Brian A., and Elizabeth L. Andress. "Resources for Starting Your Own Preserved Foods Business: Can

I Sell My Home Preserved Food?" National Center for Home Food Preservation, June 2006. Retrieved January 14, 2009 (http://www.uga.edu/nchfp/business/starting_business.html).

Peters, Louise. "Tableland." Qzine Appetizing Information, January 24, 2009. Retrieved January 25, 2009 (http://qzine.blogspot.com/2009/01/tableland.html).

Petrini, Carlo. *Slow Food Nation: Why Our Food Should Be Good, Clean, and Fair.* New York, NY: Rizzoli International Publications, 2007.

Pettigrew, Stacy. "Bad Biotech, Parts 1 and 2." Wings radio series, Department of Women's Studies, University of South Florida, 2008. Retrieved January 4, 2009 (http://web3.cas.usf.edu/main/depts/WST/wings.aspx).

Pfeiffer, Dale Allen. *Eating Fossil Fuels: Oil, Food and the Coming Crisis in Agriculture.* Gabriola Island, BC, Canada: New Society Publishers, 2006.

Pollan, Michael. "We Are What We Eat." Center for Ecoliteracy, 2005. Retrieved January 14, 2009 (http://www.ecoliteracy.org/publications/rsl/michael-pollan.html).

Redlin, Janice, ed. *Land Abuse and Soil Erosion: Understanding Global Issues.* New York, NY: Weigl Publishers, 2007.

Rogers, Stan. "The Field Behind the Plough." Northwest Passage, 2006. Retrieved January 22, 2009 (http://stanrogers.net).

Stone, Michael K. "It Changed Everything We Thought We Could Do." Center for Ecoliteracy, 2002. Retrieved January 14, 2009 (http://www.ecoliteracy.org/publications/straw.html).

Thomson, Jennifer. *Seeds for the Future: The Impact of Genetically Modified Crops on the Environment.* Ithaca, NY: Cornell University Press, 2007.

U.S. Department of Agriculture. "Frequently Asked Questions About Biotechnology." July 29, 2008. Retrieved January 14, 2009 (http://www.usda.gov/wps/portal/!ut/p/_s.7_0_A/7_0_1OB?contentidonly=true&navid=AGRICULTURE&contentid=BiotechnologyFAQs.xml).

Valentine, M. F. "The Peaceful Potato." Knittymuggins, January 1, 2009. Retrieved January 20, 2009 (http://knittymuggins.wordpress.com/2009/01/01/the-peaceful-potato).

White, D. Andrew. "Organic & Natural Pest Control." Urban Forestry & Gardening Consulting, September 9, 2008. Retrieved January 8, 2009 (http://www.ontarioprofessionals.com/organic.htm).

Index

About the Author

Paula Johanson has worked as a writer, editor, and teacher for more than twenty years. Her most recent titles for young adults are *Making Good Choices About Fair Trade* and *Processed Food*. She operated an organic method market garden for fifteen years, selling produce and sheep's wool at farmers' markets. At two or more conferences each year, she leads panel discussions on practical science and how it applies to home life and creative work. An accredited teacher, Johanson has written and edited educational materials for the Alberta Distance Learning Centre and eTraffic Solutions in Canada.

Photo Credits

Designer: Sam Zavieh; Editor: Kathy Kuhtz Campbell; Photo Researcher: Cindy Reiman